# The
# Hydrogen
# Bomb

## Unleashing the Nuclear
## Age and Arms Race

TAMRA ORR

The Rosen Publishing Group, Inc., New York

Published in 2005 by The Rosen Publishing Group, Inc.
29 East 21st Street, New York, NY 10010

**Library of Congress Cataloging-in-Publication Data**

Orr, Tamra.
The hydrogen bomb: unleashing the nuclear age and arms race / by Tamra Orr.
    p. cm.— (The library of weapons of mass destruction)
Includes bibliographical references and index.
ISBN 1-4042-0293-5 (library binding)
1.  Hydrogen bomb—History—Juvenile literature.
I. Title. II. Series.
UG1282.A8.O79 2004
623.4'5119—dc22
                                                            2004012432

*Manufactured in the United States of America*

**On the cover:** The first explosion of the thermonuclear, or hydrogen, bomb at Eniwetok Atoll known as the Mike Test, in the South Pacific on November 1, 1952, local time (October 31 in the United States).

# CONTENTS

# INTRODUCTION

On August 6, 1945, an American physicist named Luis Alvarez (1911–1988) wrote a letter to his son on his way back to the Pacific island of Tinian from Hiroshima, Japan, where the atomic bomb "Little Boy" had just been dropped from the *Enola Gay*. Alvarez was testing the explosive yield of the bomb from the back-up plane, the *Great Artiste*. He included the letter in his book *Alvarez* years later. He wrote:

What regrets I have about being a party to killing and maiming thousands of

The United States conducted 1,054 nuclear tests between July 16, 1945 (the date of the Trinity Test in New Mexico) and September 23, 1992. The only two nuclear attacks were conducted on the cities of Hiroshima and Nagasaki, Japan, which ended World War II. Tests were conducted between 1946 and 1962 in the Marshall Islands of the South Pacific, but since then, all U.S. tests have been conducted at its Nevada Test Site.

Japanese civilians this morning are tempered with the hope that this terrible weapon we have created may bring the countries of the world together and prevent further wars.

As destructive as this weapon was, there was already speculation among its creators that the atomic bomb could possibly be only one part of an even larger, more powerful explosion—the hydrogen bomb. The development of both of these weapons changed how conflicts have come to be resolved, especially as various nations have gained possession of nuclear technology. Before nuclear weaponry, military strategy was based on defense. For the United States this had not been difficult. It had no militarily powerful neighbors, and was otherwise isolated by two vast oceans, making an enemy's attack nearly impossible to sustain. But when the United States' Cold War adversary, the Soviet Union, also acquired nuclear weapon technology and both superpowers then improved the delivery systems of such weapons that included long-range bombers and ballistic missiles, the United States indeed became vulnerable for the first time. Suddenly, the best defense was to deter its enemy (in this case, the Soviet Union) from using this new destructive force by assuring the enemy of an immediate counterattack of equal destruction. In the following chapters, we will explore how the hydrogen bomb came to be the United States' response to the Soviet development of nuclear weaponry, unleashing the arms race and nuclear age for humankind. ■

"Joe 1" was the Soviet Union's first nuclear test detonated in Kazakhstan on August 29, 1949. It was a carbon copy of the "Fat Man" bomb, based on information supplied by Klaus Fuchs. The Soviet Union had considered were other bomb designs at the time, but given the tensions with the United States, Soviet scientists decided to go with what they knew would be successful.

# 1

# FROM THEORIES TO WEAPONS

The atom was only a theory until the beginning of the twentieth century. Then, in that century's first four decades, multiple theories would be tested and proven about this smallest component of all matter and how it could even be broken down further into its components: neutrons, protons, electrons, and the nucleus.

The concept of atoms had been pondered as far back as the fifth century BC by the ancient Greek Democritus (460–370 BC). However, these building blocks would not be further characterized until scientists began to make discoveries about electricity and magnetization in the seventeenth and eighteenth centuries. What immediately led to the twentieth-century discovery of the atom? The first discoveries came from experimentation with cathode rays. These rays were the product of evacuating air from a glass tube that contained metal plates and was sealed at each end. These plates were then connected to a battery or induction coil. When receiving an electrical charge, a glow began to move inside the tube from the negative plate, known as the cathode, to the positive plate, the anode. Many scientists began experimenting with the cathode ray tube.

One such scientist was a German physics professor named Wilhelm Konrad Röntgen (1845–1923). On November 8, 1895, he enclosed a cathode ray tube in a sealed black carton, and studied the effects of the rays through different gases. In the darkened room in which he worked, he noticed a glow coming from a chemically treated sheet of paper he was using with other experiments across the room. He knew the glow was originating from the tube, but that it could not be from the cathode rays itself. He labeled these strange, glowing rays "X-rays." A week later, he took the first ever X-ray picture of his wife's hand, revealing her bones and jewelry. News of these rays stunned the scientific world, sending many scientists working to learn more.

The British physicist J. J. Thomson (1856–1940), in 1897, would then discover the components of the atom that came to be known as electrons, the negatively charged particles that seemed to flow from the cathode rays, no matter which gases they were exposed to. He went on to theorize the structure of the atom and how to separate different kinds of atoms.

In France, physicist Antoine-Henri Becquerel (1852–1908), as well as Marie (1867–1934) and Pierre Curie (1859–1906), had experimented with the recently identified elements uranium and radium that were known to emit rays. They were concentrating on how these elements could be made radioactive, or capable of emitting radiation. Their findings excited the scientific world and for several decades, scientific experts were

$$\alpha = \frac{\hbar}{ec}$$

$$\frac{p^2}{m} = k_1 E.$$

$$\sqrt{m^2 c^4 + c^2 p^2} = E$$

Italian-born physicist Enrico Fermi was the first to achieve nuclear fission when he shot neutrons through paraffin, which initiated the first chain reaction in uranium. He then built the first atomic pile reactor in 1942. The scientists at Los Alamos had arranged a betting pool on the size of the Trinity Test's explosive yield before the day of the test. Fermi bet that the explosion could not be contained, and would wipe out all life on Earth.

exploring radioactivity and how it could be used. In 1934, when Italian American physicist Enrico Fermi (1901–1954) learned how to split the uranium atom, a process that came to be known as fission, he did not know that he had just ushered in a new era of

science—one that would alternately fascinate and terrify people all over the world. The energy released during the fission process was surprising and full of almost limitless potential.

By the end of the 1930s, more than 100 papers had been written about how fission could be applied in the world. Each one proved the same theory over and over: fission could result in the biggest and most powerful blast of energy the world had ever seen. By applying Albert Einstein's famous equation $E=mc^2$, a huge amount of energy could result from the atom's release of its binding energy in the fission process.

What might have come of this knowledge in a time of peace will never be known. Just as the power

Tanks fire into the night during the Battle of Stalingrad, one of the bloodiest battles in history. It lasted from August 19, 1942, until February 2, 1943. Total casualties were estimated between 1 and 2 million. The capture of the city had been strategically important for Hitler because it was a major Soviet industrial town. However, by the beginning of 1943, the German soldiers, cut off from their supplies by the Soviets, were starving. Inset: The B-17 Flying Fortress was used for strategic bombing of enemy ports and industry by the Allies in World War II.

of splitting the atom was becoming news, so were reports of a quickly approaching war. The year was 1939, and Adolf Hitler's German armies were beginning to conquer portions of Europe. One by one, cities were toppling under Hitler's unrelenting bombardment, as he sought to expand his power throughout Europe and establish the Germans as the master race of people. A large portion of victory in World War II (1939–1945) was dependent upon advancements in weaponry. Since World War I (1914–1918), the long-range bomber, the dive bomber, and the aircraft carrier had been introduced. As experiments kept proving the power of the fission process, it was not long before atoms would be the key to the most powerful arsenal among warring nations.

Could the splitting of an atom really be the key to winning a war? Hungarian physicist Leo Szilard (1898–1964) certainly thought so. He

Leo Szilard first met Albert Einstein and Max Planck at the University of Technology in Berlin. He is remembered mostly for initiating the Manhattan Project, yet after World War II, he concentrated on the control of nuclear arms and positive uses for atomic energy, finally abandoning physics for biology.

was just one of numerous eastern European scientists who had fled the discrimination against Jews that preceded the rise of Hitler's Nazi regime. Szilard was consumed with worry that the Germans would beat the United States in the race to create that first weapon, now being called the atomic bomb. In order to convince the people in power that developing a bomb as fast as possible was the best route of action, Szilard turned to one of the world's most trusted and well-known scientists, Albert Einstein (1879–1955), who had also moved to the United States as Hitler was rising to power.

Because Einstein was a pacifist, it went against his personal ethics to endorse the development of a bomb. Like Szilard, however, he was very concerned that Germany would indeed build a bomb before the United States did, and that notion was frightening. On August 2, 1939, Einstein wrote a letter directly to U.S. president Franklin Roosevelt (1882–1945). In it, he encouraged the president to speed up the production of the country's first real atomic bomb.

The letter did not reach President Roosevelt for more than two months. After he read it, he appointed the organization of the Uranium Advisory Committee and gave it a limited amount of money with which to purchase the uranium needed to continue its experiments. At first, the committee was slow in making progress. It was not until early December 1941, that the pressure of the ever-growing war convinced the government it was time to accelerate its process. Before any decision could be made, the world was stunned by Japan's surprise attack of Pearl Harbor on December 7, 1941, that killed more than 3,000 American soldiers and their families. Japan, which at this time had been at war with China in an effort to expand its power throughout Asia, and now looked to

Klaus Fuchs, Soviet physicist and spy, was nicknamed Penny-in-the-slot Fuchs, while at Los Alamos. The reserved, hardworking member of a team of scientists sent from Britain was known to speak only when spoken to. He passed along atom bomb secrets to the Soviet Union between 1942 and 1949. Fuchs had fled to Britain from Germany when his Communist political affiliations made him a target of the Nazis.

The Allied leaders at the Potsdam Conference were
Winston Churchill, Harry Truman, and Joseph Stalin.
The Big Three met in the Potsdam suburb of Berlin in
July 1945, to discuss postwar Europe and bringing an
end to the continuing war in Japan. Truman casually
mentioned the atomic bomb to Stalin, later describing
in his memoirs Stalin's lack of interest. Stalin already
knew about the Trinity Test.

control the Pacific, had allied itself with Germany and Italy—the Axis powers. The United States would now enter the war, allied with Great Britain, France, and Russia. Suddenly, the atomic bomb was catapulted from a future possibility to an immediate necessity.

## TESTING THE POSSIBILITIES

Almost a year later to the day of the Pearl Harbor attack, the first chain reaction of splitting atoms was achieved in a small test site at an abandoned athletic court at the University of Chicago. Scientists were elated and began working even harder on producing the world's first atomic bomb. Much of their work was done in a remote area of north-central New Mexico, at a former residential boys' school in Los Alamos. Despite intense security at the site, employees Klaus Fuchs (1911–1988) and David Greenglass (1922– ) leaked top-secret information about the bomb's construction and development to Russian spies. Both were later caught and sent to prison for treason, along with Greenglass's sister Ethel Rosenberg and her husband, Julius.

On April 12, 1945, President Roosevelt died and Vice President Harry S. Truman (1884–1972) was sworn in to take his place. Less than a month later, on May 7, Germany surrendered and the war in Europe was over. The fear that the Germans would be the first to create a bomb no longer haunted the experts at Los Alamos. Their focus now shifted to the ongoing war in the Pacific. Truman, attending the Potsdam Conference in July 1945, outside Berlin, hoped to gain Soviet leader Joseph Stalin's support in fighting the Japanese. Knowing that the United States had the world's most incredible weapon ready to be used if needed gave Truman the extra confidence he wanted before talking to Stalin. Truman knew that with a bomb like this at his disposal, Stalin's support in this battle would be welcomed, though no longer required, for victory.

On July 16, the time to officially test the atomic bomb arrived. The location was about 200 miles (322 km) south of Los Alamos in a desolate area called Jornada del Muerto, or "Journey of the Dead." It was a 24- by 17-mile (39 by 27 km) area. A 100-foot (30.5-meter) tower sat in the middle of it and the shed at the top was home to the bomb. Named the Trinity Test, the detonation was set for 5:30 in the morning.

Sirens blared and flares lit the sky to send out a warning. The bomb's designers, creators, mechanics, and other interested parties were gathered at a safe site more than 10 miles (16 km) away to watch the historical event.

When the bomb went off, the tower was literally vaporized and the surrounding sand melted into a green glass. A saucer-shaped crater was carved a quarter mile across the ground. The testing of the world's first atomic bomb was considered by most to be a complete success.

## HEADING FOR JAPAN

Truman was ecstatic when he got the news about how well the atomic bomb's test went. Now he no longer had to depend on the Soviet Union's support to win the war

General Leslie Groves (center), Robert Oppenheimer (*far left*, in hat), and others examine Ground Zero, the ground below where "Gadget" had sat in a 100-foot (30.5 m) tower known as Site X. The twisted wreckage, inspected here on September 11, 1945, was all that remained after the detonation of the first nuclear weapon. It was at this time that the press was finally allowed access to the Trinity Site, mainly in an effort to dispel radioactivity rumors, making front-page news across the country.

against Japan. Instead, his country had what he considered a true ace in the hole. On July 26, the United States, along with Great Britain and China, sent an ultimatum to Japan. It was to surrender immediately or be completely destroyed.

Japan refused this ultimatum over a radio broadcast, stating that the terms of surrender were simply not acceptable. President Truman was not surprised, but he was disappointed. If the United States went with the plan of a massive invasion of Japan, the estimates of casualties on both sides were expected to surpass the casualties caused by dropping the bomb, if the bomb worked as advertised. (There were those high government officials who remained skeptical, even after the successful Trinity Test.) Truman, however, thought the decision was inevitable. It was time to drop the atomic bomb

> **"We had adopted an ethical standard common to the barbarians of the Dark Ages."**
>
> *Admiral William D. Leahy*

On August 6, 1945, at 8:15 AM, "Little Boy" was dropped on the Japanese city of Hiroshima. More than 66,000 people were killed instantly. Another 69,000 were injured and many died days to weeks later. When Japan still refused to surrender, a second bomb named "Fat Man" was dropped three days later on the city of Nagasaki. This time, almost 40,000 people were killed and 25,000 were injured. Later, thousands more would die from wounds and radiation illnesses such as leukemia. Finally, Emperor Hirohito admitted defeat in August 1945, and surrendered. At last, World War II was completely over.

Many of the scientists and others involved in creating the atomic bomb were horrified at the massive casualties from the attacks. It was one thing to put theories and mathematical equations on blueprints and reports, but another to see that same information used to destroy entire cities. Some of the Los Alamos scientists walked away and never returned to working on weaponry; others shifted to positive aspects of the science, including research into atomic energy as a source of electricity.

Even the president's closest advisers had deep regrets about the use of the atomic bomb in warfare. Admiral William D. Leahy (1875–1959), who had served as chairman of Truman's Joint Chiefs of Staff, the executive agency that advises the president on military questions, had this to say about the dropping of the atomic bombs in his autobiography, *I Was There*:

> It is my opinion that the use of this barbarous weapon at Hiroshima and Nagasaki was of no material assistance in our war against Japan. The Japanese were already defeated and ready to surrender . . . My own feeling was that in being the first to use it, we had adopted an ethical standard common to the barbarians of the Dark Ages. I was taught not to make war in that fashion, and wars cannot be won by destroying women and children.

Of course, there were those that remained dedicated to creating bombs, such as Hungarian mathematician Edward Teller (1908–2003). Teller was Jewish and had been able to escape Hitler's rise to power in Europe, eventually seeking refuge in the United States. He was one of the biggest supporters of the atomic bomb, and until his death in 2003, he would continue to support the nation's creation of bombs. Teller and many of his colleagues believed these destructive weapons would help protect the nation and keep it one of the most powerful countries in the world. These men, along with a Polish American mathematician named Stanislaw Ulam (1909–1984) looked beyond the atomic bomb to the potential of

# TIMELINE OF THE 1950s AND 1960s

**November 1, 1952**  The United States explodes the first H-bomb

**August 12, 1953**  The Soviet Union explodes its first H-bomb

**March 1, 1954**  The United States tests massive H-bomb at Bikini

**May 21, 1956**  The United States drops the first H-bomb from a plane

**May 15, 1957**  Britain drops its first H-bomb

**June 17, 1967**  China detonates its first H-bomb

(http://www.foxnews.com/printer_friendly_story/0,3566,76888,00.html)

This picture of Hiroshima was taken one month after the atom bomb was dropped on the city, detonating at 1,850 feet (564 m) above, and killing between 70,000 and 130,000 people. The buildings left had been earthquake reinforced. Today, the explosive yield of the U.S.'s Minuteman III missile is equal to that of more than eighty Hiroshima-strength bombs.

creating something even bigger and more powerful. After all, the atomic bombs brought about the end of the war, so they were clearly effective tools. What else was out there? What other possibilities existed? How big could a bomb get?

## GOING THERMONUCLEAR

It might have been years before those questions were answered; it might never have happened at all except that the U.S. military soon collected solid information that the Soviet Union had not only caught up with U.S. technology about the atomic bomb, but also had already tested one of its own. Although the United States and the Soviet Union had put aside their differences to fight their common enemy, Germany, in World War II, tensions seemed to restart immediately afterward between the two superpowers in what would become known as the Cold War, lasting until 1991. Now, the arms race would begin.

On August 29, 1949, the Soviet Union tested "Joe 1," which, at 20 kilotons, made it the same size as the Trinity Test bomb. President Truman then made a public announcement on September 23, informing the American people that such an explosion had occurred in the USSR. Americans were shocked. The images of what had happened in Japan were still fresh in Americans' minds and they were not sure how to respond to this unexpected threat. At the time of this announcement, however, President Truman had not been made aware of the possibility of the hydrogen bomb. Yet after being briefed on the possibility of this bomb in early October, the course for the future became clear to him.

On January 22, 1950, the president announced to America that work on the hydrogen bomb, or super bomb, as it was now being called, would forge ahead immediately. Unlike the atomic bomb, which works through a process of fission, this bomb would use fusion and would be a thermonuclear bomb. The atomic bomb is based on the splitting of large atoms of uranium or plutonium, the heavier elements of the periodic table. However, the hydrogen bomb (H-bomb) is based on small atoms of hydrogen with its low atomic weight, fusing with the nucleus of the next lightest element in the periodic table, helium. This action requires an enormous amount of heat to work, comparable to the heat of the Sun. The fusion process is more than 1,000 times more powerful than fission. It creates enormous shock waves, powerful winds, and deadly radiation.

However, within the Atomic Energy Commission (AEC) the president's decision caused much debate. This commission, established in 1946 by Congress, transferred the control of the use of atomic energy away from the military and into civilian hands, in an effort to promote the international use of atomic energy for peaceful means. Edward Teller and Lewis Strauss (1896–1974) of the commission strongly supported Truman. Yet scientists such as Enrico Fermi just as strongly opposed making a super bomb. In a joint statement to the president made with Austrian physicist I. I. Rabi (1898–1988), the physicists described how, because no limits existed on the weapon's destructiveness, its very existence was a danger to all of humanity. Even Einstein spoke out, writing that the radioactive poisoning of the atmosphere caused by the H-bomb could lead to the complete destruction of life on Earth. While many campaigned for just increasing the production of atomic bombs rather than attempting to build something bigger, their voices were ignored and lost in the rush of excitement over the potential of this new H-bomb.

More complex and sophisticated, this super bomb was also far more destructive. Whereas the explosive power of the atom bomb was roughly equivalent to 20,000 tons (20 kilotons) of TNT, the hydrogen bomb was measured in megatons, equivalent to millions of tons of TNT. Indeed, the atomic bomb was thousands of times more powerful than TNT, and the hydrogen bomb was thousands of times more powerful than the atomic bomb. As the decade came to an end, interest in it escalated and with it came controversy, debate, and raging ambivalence on whether it was going to be the nation's guardian or its eventual destructor. ■

# 2

# CONTROVERSY
# AND
# CONFUSION

All was not calm and composed within the Atomic Energy Commission. The members felt pressured by the president, the nation's people, and the scientists of the commission to make some kind of firm decision about whether to go ahead with the development of the H-bomb. In October 1949, they turned to the General Advisory Committee, chaired by Robert Oppenheimer (1904–1967), the American physicist who was

J. Robert Oppenheimer served as chairman of the Atomic Energy Commission from 1943 until 1953. His lack of support of development for the hydrogen bomb, along with his political background that had at one time included Communist sympathies, led to the revocation of his security clearance, ending his government work.

the scientific director of the Manhattan Project, the team of scientists who designed the atomic bomb and tested the bomb at Los Alamos. The Atomic Energy Commission needed advice on what to do. It believed that the developing of the H-bomb was the best route to go, but wanted feedback from other experts. It was not an easy decision for the General Advisory Committee either. For two days, its members debated the issue and finally agreed that uranium and plutonium stores should continue to be brought in for use in the production of multiple atomic bombs. Regarding the new super bomb, they filed an official report that stated they were not in support of the idea simply because such a weapon would not be easily controlled and could lead to catastrophe.

An island that is part of Enewetak Atoll, in the South Pacific, is pictured here after the Mike Shot was detonated on October 31, 1952, in the United States (November 1, local time). The atoll, located 3,000 miles (4,828 km) west of Hawaii, is a ring of forty small islands. It was captured by the United States from the Japanese in 1944. After being designated as a testing site for nuclear weapons, all of its native inhabitants were then moved to other locations.

The Atomic Energy Commission succeeded the Manhattan Project on January 1, 1947. This marked the end of Leslie Groves's position as director. This agency, which then controlled the development of atomic energy, was the predecessor of the U.S. Department of Energy. David E. Lilienthal *(seated, left)* was chosen by Harry Truman as the commission's chairman, formerly having been a lawyer who had served as head of the Tennessee Valley Authority.

They unanimously recommended that the United States not embark on any program to build the H-bomb.

The General Advisory Committee's official report stated that there was "no limit to the explosive power of the [hydrogen] bomb itself except that imposed by requirements of delivery." Its conclusion was that, "it is not a weapon which can be used exclusively for the destruction of material installations of military or semi-military purposes. Its use, therefore, carries much further than the atomic bomb itself the policy of exterminating civilian populations."

The Atomic Energy Commission was not pleased with this recommendation. There was division even within this organization, with some officials for and others against the development of this weapon. Chairman David Lilienthal (1899–1981) wanted to speed up the production of atomic bombs, but Commissioner Strauss did not agree. On November 25, 1949, he wrote directly to President Truman stating his beliefs. "I believe that the United States must be as completely armed as any possible enemy," he wrote. "From this, it follows that I believe it unwise to renounce, unilaterally, any weapon which an enemy can reasonably be expected to possess. I recommend that the President direct the Atomic Energy Commission to proceed with the development of the thermonuclear bomb." The Joint Chiefs of Staff agreed with Strauss's statement and said, "The United States would be in an intolerable position, if a possible enemy possessed the bomb and the United States did not."

> "The hydrogen bomb controversy marked the first time that a large group of scientists argued for remaining ignorant of technical possibilities."
>
> *Edward Teller*

Truman continued to struggle with what to do. With so many of his experts disagreeing with each other and making different recommendations, it was a difficult time. As Teller wrote in *Memoirs: A Twentieth-Century Journey in Science and Politics*, "The hydrogen bomb controversy marked the first time that a large group of scientists argued for remaining ignorant of technical possibilities." Finally, Truman turned to a special committee of the National Security Council for its advice. The council included Secretary of State Dean Acheson, Secretary of Defense Louis Johnson, and Lilienthal. They agreed that pursuing the new super bomb was the wisest course of action and in the end, the bottom line statement was that the AEC "investigate the feasibility of the H-bomb."

On January 31, 1950, President Harry Truman made the following statement to the American public regarding development of the hydrogen bomb:

It is my responsibility as Commander in Chief of the Armed Forces to see to it that our country is able to defend itself against any possible aggressor. Accordingly, I have directed the Atomic Energy Commission to continue work on all forms of atomic weapons, including the so-called hydrogen or super bomb. Like all other work in the field of atomic weapons, it is being and will be carried forward on a basis consistent with overall objectives of our program for peace and security.

This we shall continue to do until a satisfactory plan for international control of atomic energy is achieved. We shall also continue to examine all those factors that affect our program for peace and this country's security.

## RESPONSE FROM THE USSR

The Soviet Union's response to the president's announcement was immediate. Four days after the message was broadcast, Lavrenty Beria, the former deputy to the chief of the Soviet secret police, who was known for his cruelty, ordered his scientists to give him a detailed report on the progress of their "layer cake" bomb. This was physicist Andrey Sakharov's (1921–1989) design for a hydrogen bomb that consisted of alternating layers of light elements (such as tritium and deuterium) and heavy elements (U-238, an isotope of uranium). Although Sakharov would one day be known for his dedication to peace and human rights, he was also called the father of the Soviet hydrogen bomb. His first brush with the concept of powerful bombs was when the United States dropped the atomic bomb on Hiroshima. In his memoir, he described the moment: "I was so stunned that my legs practically gave way," he wrote. "There could be no doubt that my fate and the fate of many others, perhaps of the entire world, had changed overnight. Something new and awesome had entered our lives, a product of the greatest of the sciences, of the discipline I revered." He was recruited to work on the Soviet nuclear bomb in the summer of 1948.

After announcing to the country that the hydrogen bomb program would be expanded, Truman secretly instructed the AEC to expand its facilities in preparation for the H-bomb's production. Oppenheimer and others objected to Truman's decision. They thought it would be better if the United States went to the Soviet Union with a proposal that they both agree to stop producing weapons. They also strongly endorsed telling the people about the ramifications of using a bomb as powerful as the H-bomb. Both of their suggestions were ignored.

## A NEW DESIGN

Edward Teller worked hard on designing a functional hydrogen bomb, but kept running into problems. Just when he thought he had it figured out, Stanislaw Ulam showed Teller that it would not work and set out to design a better one. This did not make Teller happy and there was often a lot of friction between the two men. Ulam was the first to recognize that it was possible to place each one of the H-bomb's components into a single casing. He was aware of what he was creating, later describing how such calculations that can be scribbled on a blackboard or paper can change the course of human history. He then put a fission or atomic bomb at one end and the thermonuclear material on the other. The shock waves of the atomic bomb would compress and then detonate the fusion fuel. Teller reluctantly accepted this design, only altering it so that instead of shock waves, the bomb exploded as a result of a process called radiation implosion. It was a successful design that was used in the production of all future H-bombs.

Polish-born Stanislaw Ulam became part of the team at Los Alamos in 1943. There, the mathematician developed the implosion igniter for the atomic bomb. When he reviewed Teller's calculations for the hydrogen bomb, he went about making corrections, ultimately sharing the credit with Teller as a developer of the thermonuclear fusion bomb.

Teller's conflicts extended beyond those with Ulam. He was also struggling with longtime colleague Robert Oppenheimer, whom he did not trust because Oppenheimer was so adamantly against the construction of the H-bomb. In 1952, Teller was secretly interviewed by the FBI about his associate. The FBI had discovered that Oppenheimer had some suspicious associations with Communists since his wife, brother, and sister-in-law were all followers of Communism. In his testimony to the FBI in 1952, later made public under the Freedom of Information Act in 1977, Teller stated, "I feel that I would like to see the vital interests of this country in hands which I understand better, and therefore trust more." This statement resulted in Oppenheimer having his security clearance revoked and spurred one of history's biggest feuds.

## OFF TO BIKINI ATOLL

In the central Pacific Ocean are the twenty-nine atolls and five islands that make up the Marshall Islands. Between 1946 and 1958, a total of sixty-seven nuclear tests were conducted there. On Eniwetok Atoll, the first hydrogen bomb was dropped on November 1, 1952. Called the Mike Test, it was 500 times more powerful than the atomic bombs used in Japan. It completely destroyed the island. The following year, the USSR detonated its first hydrogen bomb. More bombs were dropped and tested by the United States in a nearby atoll called Bikini. The 167 residents of the island were evacuated and many would not ever be able to return to their homes because of high levels of radioactivity.

On March 1, 1954, the largest hydrogen bomb was tested on Bikini Atoll. On this small island in the Pacific, the Bravo bomb was dropped and its explosion not only blew out the measuring instruments, but also surpassed the scientists's expections who had created it. While they thought it would have a blast equivalent to 5 million tons of TNT, it actually far surpassed that, reaching 15 megatons, a thousand times more powerful than the atomic bomb dropped on Hiroshima just a few years earlier.

The Baker shot, detonated on July 15, 1946, was part of Operation Crossroads at Bikini Atoll. Whereas the Trinity Test was shot to study the behavior of the nuclear weapons' design, Baker was part of the first weapons effects test. A fleet of seventy-one ships was anchored nearby and used as targets. Baker was detonated from 90 feet (27 m) below the surface and produced 20,000 tons of TNT. It was the Baker shot, the final test of the Manhattan Project, that introduced radioactive fallout.

This bomb created a humongous fireball more than 3 miles (5 km) in diameter, followed by a mushroom cloud that reached 20 miles (32 km) up in the sky. The blast could be seen for more than 135 miles (217 km) away. Winds as strong as those that accompany hurricanes stripped the remaining trees on the island. Natives on nearby islands vividly remember the moment. Their descriptions were reported around the world by news agencies. Lemyo Enob, who was fourteen at the time, stated, "It was the first time I saw the sun rise in the west."

# HOW THE HYDROGEN BOMB DESTROYS

The hydrogen bomb does damage in many ways. Initially, there is the blast, or the explosive force, which comes from the center of the explosion. The change in the air pressure caused by the blast from a hydrogen bomb destroys buildings. The accompanying strong winds destroy people, trees, and objects. Second, there is fire ignited by thermal heat. The heat of a thermonuclear blast is similar to the temperature of the surface of the sun. Then there is the effect of radiation. Gamma radiation that emits from the fission process of the hydrogen bomb can travel many centimeters into human tissue and organs. Finally, another effect is that of nuclear fallout. The explosion creates enormous quantities of radioactive dust and dirt (which make up the famous mushroom-shaped cloud) that can be carried by the wind for thousands of miles before falling to the ground, harming whoever comes in contact with it.

Any single nuclear bomb can produce some scale of these effects, depending on the size and purpose of that particular weapon. However, the most horrifying possibility would be the detonation of hundreds of thousands of these thermonuclear bombs. This would produce enormous fires and smoke, fundamentally changing the climate of Earth, bringing about what is known as a "nuclear winter."

A mile-wide crater had been blasted into the reef and physicist Marshall Rosenbluth, on a ship 30 miles (48 km) out, described the fireball, saying it "just kept rising and rising, and spreading . . . And the air started getting filled with this gray stuff, which I guess was somewhat radioactive coral."

About ninety minutes after the initial explosion, a strange kind of snowlike element began falling from the sky. A Japanese fishing vessel called the *Lucky Dragon* was about 80 miles (129 km) away from Bikini when the flakes began to rain down. On board were twenty-three fishermen who had no idea what was happening. By the time they returned to port two weeks later, all of them were ill. The radioactive fallout came to be known as the ashes of death.

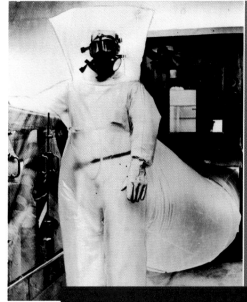

The fishermen were not the only casualties. People on nearby islands found odd ashes falling in their water and food. Children played in the powder. Some natives were evacuated within hours; for others it was a day or more. By that time, many were covered in burns and losing their hair from the radiation poisoning.

How did this happen? In a press conference held shortly after the explosion, AEC commissioner Lewis Strauss claimed it was because of the weather. He stated that "meteorologists had predicted a wind condition which should have carried the

A radiation suit, being worn here by an AEC worker in a plutonium plant, allows the extent of contamination to be examined in radioactive, or "hot" areas. In recent years, numerous employees of the AEC and Department of Energy have suffered from the effects of working in such radioactive environments.

fallout to the north of a group of small atolls lying to the east of Bikini . . . The wind failed to follow the predictions but shifted south of that line and the little islands of Rongelap, Rongerik, and Utirik were in the edge of the path of the fallout."

> "It was the first time I saw the sun rise in the West."
>
> *Lemyo Enob*

This statement may not have been completely honest. First, an earlier weather report did indeed point out the shift in the winds. Second, and perhaps most important, the reason innocent people had been exposed to these dangers was simply that the blast was far bigger than anyone had ever expected.

The following year, the United States paid the *Lucky Dragon*'s crew members and their families $2 million in damages, but that was not the end of it. Since then, more than $63,127,000 has been awarded to more than 1,500 people injured in these nuclear blasts. Some cases are still pending even today. Presently, one of the greatest afflictions passed on to the offspring of survivors of exposure to radiation is the blood disease leukemia.

In 1957, Britain dropped its first H-bomb. Ten years later, China did the same. The thermonuclear bomb technology reached across the globe now, and instead of feeling powerful and protected, people around the world were feeling frightened. As awareness spread of what devastation and destruction these bombs could create, objections and fears skyrocketed. ■

A sign indicating a nearby fallout shelter. The Civil Defense Department passed a measure to survey basements, tunnels, and other areas where civilians could take refuge in case of a nuclear war, in order to shield themselves against the harmful gamma rays that accompany radioactive fallout. These shelters were usually stocked with food paid for by the federal government.

CAPACITY 1730

3

# THE FEARS
# AND THE
# FACES

In an attempt to soothe the growing alarm of Americans, a lighthearted campaign was created around the whole idea of the bombs. The Washington Press Club had an alcoholic drink called the atomic cocktail. Disney produced an animated feature for kids called *Our Friend the Atom*, showing all the friendly ways that atomic energy might be used in futuristic communities. This

film showed how one day there would be atomic airplanes. Other public campaigns showed the use of smaller versions of the H-bomb to help move mountains for highway construction or to redesign landscape for new housing additions. Comic-book characters such as the Incredible Hulk showed how radioactive exposure could give a person immense and positive powers. The character Bert the Turtle was featured on government posters advising the American people to remember to "duck and cover." In case of a bomb attack, a person was supposed to drop to the ground and seek cover, as portrayed by Bert. Kids in school were put through regular drills of hiding under their desks at the sound of the air raid siren. Even the fashion industry tried to lend a hand by creating a very tiny new kind of swimsuit called a bikini,

Bert the Turtle taught schoolchildren how to duck and cover in 1950. When the United States monopoly on nuclear weapons was broken by the Soviet Union, campaigns were launched to alert the public of the nuclear threat. This campaign, at least as far as nuclear attacks, was quickly considered obsolete, as no one would be able to survive a nuclear blast near its center. However, it was helpful for the threat of hurricanes and tornadoes.

named after the island where many of the bombs were tested. The suit was said to have an "explosive effect."

Hollywood had its say, too. A movie called *Dr. Strangelove, or How I Learned to Stop Worrying and Love the Bomb* was a dark comedy released in 1964. The main character was supposedly based on Edward Teller. The government issued civil defense pamphlets and films for citizens to watch and see how they could lessen the risks of a bomb attack. The government media strategically walked the line between downplaying the danger and making it clear that some precautions had to be taken. For example, although dangers of radioactive ash, or fallout, would be described in the pamphlets, it would also be treated as a type of "dust" that one could get rid of simply by taking a shower or dusting off the tops of cans and jars in the kitchen.

Despite these efforts, the American people were still very frightened. They were learning all about the effects of fallout and radiation, and they knew that simply ducking and covering would not begin to protect them. People began building underground bomb shelters stocked with weeks or even months' supplies of food, water, batteries, auxiliary lights, and first-aid equipment. The Interstate Highway Act was established in 1956 to enable all states to build major highways and freeways. This would help Americans evacuate the major cities if needed.

In 1953, Dwight D. Eisenhower (1890–1969) was elected president of the United States. When asked about the use of nuclear weapons, he stated, "You can't have this kind of war. There just aren't enough bulldozers to scrape the bodies off the streets."

At the same time, the Atomic Energy Commission began developing nuclear power plants. These stations consisted of nuclear reactors where nuclear fission chain reactions are initiated and controlled at a steady rate, producing heat that can be used to generate electricity. The first civilian nuclear power unit went online in July 1957, in Santa Susana, California, and the first full-scale nuclear power plant in the country was constructed in Shippingport, Pennsylvania, that same year.

By 1960, there were an estimated 1 million family fallout shelters nationwide, according to the Department of Civil Defense. The one pictured here was designed by a privately owned atomic research company. This self-contained unit was equipped with its own gasoline generator and could sustain a large or small family, with no outside assistance, for three to five days.

## GROWING CONCERNS

As nuclear power plants continued to pop up in various places across the United States, some scientists and experts were already walking away from the construction of bombs. The truth of what they were creating was simply too disturbing. "When you see all of this yourself, something in you changes," said Andrey Sakharov. "When you see the burned birds who are withering on the scorched steppe, when you see how the shock wave blows away buildings like houses of cards, when you feel the reek of splintered bricks, when you sense melted glass, you immediately think of times of war . . . All of this triggers an irrational yet very strong emotional impact. How not to start thinking of one's responsibility at this point?"

In the early 1960s, worries about the various nations building and stockpiling weapons with

Soviet-made missiles are paraded through the streets of Havana, Cuba, on January 1, 1962, the third anniversary of the revolution that brought Fidel Castro to power in that country. At the time, Cuba made no secret of its dependence on Soviet missiles, but when it decided, several months later, to install Soviet nuclear weapons, it did so secretly.

the potential to destroy entire cities in one attack began to build. People were complaining to their government representatives. Organizations were being formed and protests were being held. By 1964, there were five nuclear weapon states: the United States, the Soviet Union, the United Kingdom, France, and China. Predictions stated that the number could reach to thirty nations in less than twenty years. President John F. Kennedy (1917–1963), who was elected in 1961, also spoke out against the use of these massive weapons. "A full scale nuclear exchange, lasting less than 60 minutes . . . could wipe out more than 300 million Americans, Europeans, and Russians, as well as untold numbers elsewhere," he said to the nation in July 1963. However, less than a year earlier, Kennedy had come as close to actually using the H-bomb as the nation ever got. When spy photographs caught the establishment of Soviet missiles in Cuba in October 1962, Kennedy ordered a naval blockade of Cuba. He told American seamen to sink the enemy's ships if they did not stop, and that would mean an act of war. American nuclear forces were put on the highest possible level of alert, but fortunately, the Soviets turned their ships around and removed their missiles from Cuba. It was a terribly tense and frightening time in American history—but it could have been far, far worse if the country had been forced to deploy nuclear warfare.

As a result of pressure from the people and politicians from all over the world, more than 180 nations met and negotiated limitations on the testing, producing, distributing and deploying of nuclear weapons. They promised to limit how many nuclear weapons were made, sold, and used. On July 1, 1968, many of them signed the Nuclear Non-Proliferation Treaty, which remains one of the most important arms control treaties in history.

## DIFFERENT THREATS

As nuclear power plants continued to be built, problems began to occur and people also realized that in addition to the dangers of bombs, nuclear energy also had risks. In 1957, a plant in Liverpool, England, had

In April 1957, Japanese artists in the city of Osaka created a huge piece of traditional Japanese calligraphy calling for an end to the nuclear arms race. Perhaps because Japan is the only nation that has ever had a nuclear weapon used against it by a foreign enemy, its people have traditionally been quick to protest against the proliferation of nuclear weapons.

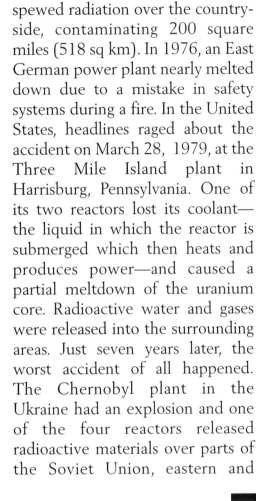

a fire in one of its reactors and spewed radiation over the countryside, contaminating 200 square miles (518 sq km). In 1976, an East German power plant nearly melted down due to a mistake in safety systems during a fire. In the United States, headlines raged about the accident on March 28, 1979, at the Three Mile Island plant in Harrisburg, Pennsylvania. One of its two reactors lost its coolant—the liquid in which the reactor is submerged which then heats and produces power—and caused a partial meltdown of the uranium core. Radioactive water and gases were released into the surrounding areas. Just seven years later, the worst accident of all happened. The Chernobyl plant in the Ukraine had an explosion and one of the four reactors released radioactive materials over parts of the Soviet Union, eastern and

The explosion of the reactor at the Chernobyl nuclear power plant *(left)* in the Ukraine in April 1986, remains the most severe nuclear accident to date. The sickly brown branches of this pine sapling *(inset)* photographed near the site in November 2000, testify to the ongoing long-term environmental damage caused by the catastrophe.

President Richard Nixon of the United States *(left)* shakes hands with Premier Leonid Brezhnev of the Soviet Union upon the signing of the SALT II Treaty in Moscow in May 1972, as an array of Soviet dignitaries looks on. This meeting was considered one of the most important summits of the two superpowers since the Potsdam Conference in 1945.

western Europe, and Scandinavia. Following each of these accidents was an outcry that this new technology was too frightening to keep using. Following the Three Mile Island accident, the U.S. nuclear energy industry created the Institute of Nuclear Power Operations to address and monitor the issues of safety and performance in power plants.

## A FAMILIAR VOICE

With the development of delivery systems of nuclear bombs moving away from aircraft, which could be shot down, to ballistic missiles, which are powered by rockets, there seemed to be no possible way to defend against such an attack. In the midst of much of this concern, a familiar figure returned: Edward Teller. In 1960, while teaching physics at the University of California, Berkeley, he had developed the concept that the best possible defense against other nation's nuclear weapons was to destroy them before they could hit the targets in the United States. Teller would then support President Richard Nixon's Safeguard Program, which set out, as Teller had described, to intercept any incoming missiles before they could land.

In 1969, the first Strategic Arms Limitation Talks (SALT) took place between the Soviet Union and the United States. On May 26, 1972, Nixon and Soviet general secretary Leonid Brezhnev signed the treaty that prohibited a nationwide missile defense, but still allowed each country two antiballistic missiles installations. In July 1974, that treaty was amended to allow only one installation.

Teller was among many who now began to argue that the United States had indeed won the Cold War against Russia without the need

# FAST NUCLEAR FACTS

**28,800** The total number of intact nuclear warheads retained by the United States and Russia

**30,000** The number of intact nuclear warheads throughout the world (17,500 are considered operational)

**128,000+** The estimated number of nuclear warheads built worldwide since 1945. All but 2 percent were built by either the United States or Russia

**10,729** The total number of intact U.S. nuclear warheads (274 are awaiting dismantlement)

**10,455** The total number of warheads in the U.S. stockpile

**8,400** Total number of operational nuclear warheads in Russian arsenal

**$3.5 trillion** The amount the United States spent between 1940 and 1995 to prepare to fight a nuclear war

**$27 billion** The amount the United States spends annually to prepare to fight a nuclear war

(From the Center for Defense Information, February 2003. http://www.cdi.org/nuclear/facts-at-a-glance.cfm)

for bloodshed, and that the victory had been made possible by the incredible advancements in science and technology in which he took part. By being in possession of such a devastating weapon yet avoiding misuse of it, death and destruction had been avoided, and would continue to be avoided by the coming generations.

As the years passed and more and more restrictions were being placed on the production and use of nuclear weapons, the fear of these weapons began to fade away. As the 1980s approached, however, it was about to return. ■

A Trident nuclear missile breaks through the surface of the Atlantic Ocean on a test flight in April 1985. The missile was fired from below the surface by a U.S. nuclear submarine. Only the United States has submarines capable of carrying and firing nuclear weapons from virtually anywhere in the world's oceans.

4

# SHIFTS IN NUCLEAR CONCERNS

The 1980s ushered in a subtle but renewed focus on nuclear bombs. For some time, the United States had been informed that the Soviet Union's weapons had grown in both quality and quantity. President Ronald Reagan (1911–2004), who held the office for almost the entire decade, was quite concerned about the information he was being given.

Premier Mikhail Gorbachev of the Soviet Union *(left)* and President Ronald Reagan of the United States *(right)* appear to be more interested in what their aides have to say than in each other at the conclusion of the summit conference between the two leaders held in Geneva, Switzerland, in November 1985.

His predecessor, Jimmy Carter (1924– ), had spoken out against nuclear weapons during his term. In his farewell address to the American people in January 1981, Carter stated, "In an all-out nuclear war, more destructive power than in all of World War II would be unleashed every second during the long afternoon it would take for all the missiles and bombs to fall. A World War II every second—more people killed in the first few hours than all the wars of history put together," he continued. "The survivors, if any, would live in despair amid the poisoned ruins of a civilization that had committed suicide." Once again Teller spoke out on the topic. He described how nuclear war had always been a real possibility for him and how he did not doubt that with the improvements the Russians had made to the bomb, the United States would be obliterated if a war were to break out. Not long after that, he told the National Press Club, "For a quarter of a century we have conceived of our situation as a balance of terror, and the dreadful point is that the terror is obvious; the balance is not." Teller

# TIMELINE OF THE 1970s TO TODAY

**March 31, 1972**  Campaign for Nuclear Disarmament's (CND) first march against nuclear weapons

**October 3, 1972**  Strategic Arms Limitation Treaty I signed by Nixon and Andrey Gromyko to freeze missile levels

**April 8, 1978**  Carter delays neutron bomb production

**June 18, 1979**  Strategic Arms Limitation Treaty II signed by Carter and Brezhnev

**June 17, 1980**  Government announces missile sites

**October 24, 1981**  CND rally attracts thousands

**March 23, 1983**  Reagan announces his "Star Wars" program to the American public

**April 1, 1983**  Demonstrators form a human chain linking nuclear sites in England

**October 22, 1983**  CND march attracts biggest crowd ever

**November 21, 1985**  Superpowers aim for "safer world"

**September 18, 1987**  Superpower treaty to scrap warheads

**December 8, 1987**  Superpowers agree to reverse arms race

**July 31, 1991**  START I signed by President George H. W. Bush and Gorbachev to cut warheads

**January 3, 1993**  START II signed by President Bush and Yeltsin for further cuts

**May 28, 1998**  World fury at Pakistan's nuclear tests

**May 24, 2002**  Treaty of Moscow signed by President George W. Bush and Putin

(http://news.bbc.co.uk/onthisday/hi/dates/stories/march/31/newsid_2530000/2530839.stm)

*Workers salvage equipment from a Titan II missile silo at Little Rock Air Force Base in Jacksonville, Arkansas, in 1987. The U.S. nuclear arsenal is spread out among many locations across the country.*

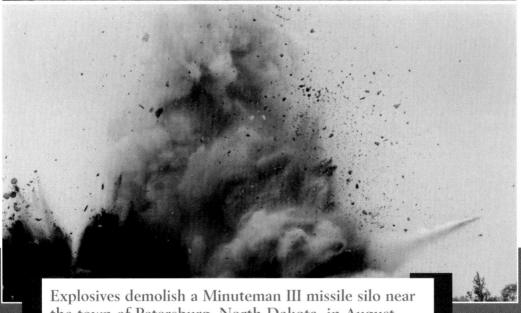

Explosives demolish a Minuteman III missile silo near the town of Petersburg, North Dakota, in August 2001. The missile site was destroyed in compliance with the terms of the first Strategic Arms Reduction Treaty (START I), concluded by the United States and the Soviet Union in July 1991. A missile silo is the underground housing in which a guided missile is kept in readiness for firing.

worked, along with many others, to convince Reagan that some kind of new safeguard was needed to protect the nation from Russian bombs.

In a nationally televised address in 1983, Reagan told the country that he planned to establish a national missile defense system, nicknamed Star Wars by its many skeptics. This system would use space-based weapons to intercept enemy missiles and, as Reagan described, make nuclear weapons impotent and obsolete. It was an extremely expensive concept and one that not everyone agreed with either. Establishing this system was a sore point between Reagan and Soviet president Mikhail Gorbachev, causing several stalls in their peace talks. As Reagan's second term came to an end, it was proposed that the Star Wars system be revamped into a type of limited system that would protect against accidental and unauthorized launches. In 1989, as President George H. W. Bush (1924– ) was sworn into office, he shifted the program to one that was based on a global early warning and tracking system.

## THE SITUATION TODAY

Currently, the United States, Russia, the United Kingdom, France, and China admit to still having nuclear weapons. South Africa is said to have built some, but has since dismantled them, as have the Ukraine, Belarus, and Kazakhstan, former parts of the Soviet Union. Israel is considered to be a threshold state, meaning it doesn't admit to having any bombs, but it is generally accepted that it either does or is able to easily get them. Pakistan and India have declared themselves nuclear states. Both have tested atomic bombs of the fission type. India may have the ability to produce a hydrogen bomb. Recently, North Korea has resumed its nuclear weapons research, despite protests from the United States, China, Russia, Japan, and South Korea. Iran is also believed to have nuclear weapon construction programs.

Nuclear power plants are still found all over the globe. The United States has 103. Other countries, including France and Japan, invest more into this technology than the United States currently does. The United Nations' International Atomic Energy Agency, located in Vienna, Austria, supervises more than 900 of these plants.

Germans celebrate the end of the Cold War atop and around the Berlin Wall, which had divided the city of Berlin between its Democratic and Communist sectors since the early 1960s. For many people, the end of the Cold War raised hope that a significant reduction in the world's nuclear weapons arsenal could be achieved, but the first years of the twenty-first century have been a time of increased concern about weapons of mass destruction.

Despite all of the research, development, controversy, and concern to date, the hydrogen bomb, or thermonuclear bomb, has never been used in anything other than test runs. Just as the atomic bomb was the stepping-stone for the hydrogen bomb, however, so this bomb may well serve as the forerunner of the new neutron, or N-bomb. Considered to be a "clean" bomb in that it has less radioactive fallout than its predecessors, this new bomb will be lethal to all living creatures, but will not harm any structures or buildings, as it works by killing via radiation. So far, the N-bomb is only a theory and many people hope that it will remain only an idea on paper. ■

# [GLOSSARY]

**atoll**  A ringlike coral island and reef that nearly or entirely encloses a lagoon.

**fission**  A nuclear reaction in which an atomic nucleus, especially a heavy nucleus such as an isotope of uranium, splits into fragments.

**friction**  Conflict, as between persons having dissimilar ideas or interests; clash.

**fusion**  The joining of atomic nuclei.

**horrific**  Causing terror, horrifying.

**isotope**  Atoms of the same element that have different masses.

**manipulate**  To alter or change, especially to serve one's purpose.

**pacifist**  A person who does not believe in using violence.

**radiation implosion** A violent inward collapse of radiation.

**radioactive**  Capable of emitting radiation.

**ramifications**  Results or consequences.

**surpass**  To go beyond, pass up.

**TNT [trinitrotoluene]**  A widely used conventional explosive, equivalent to dynamite.

**treason**  A violation of allegiance toward one's country.

# [ FOR MORE INFORMATION ]

Campaign for Nuclear Disarmament
162 Holloway Road
London, N7 8DQ
Web site: http://www.cnduk.org

Nuclear Age Peace Foundation
PMB 121
1187 Coast Village Road, Suite 1
Santa Barbara, CA 93108
(805) 965-3443
e-mail: wagingpeace@napf.org
Web site: http://www.nuclearfiles.org

## WEB SITES

Due to the changing nature of Internet links, the Rosen Publishing Group, Inc., has developed an online list of Web sites related to the subject of this book. This site is updated regularly. Please use this link to access the list:

http://www.rosenlinks.com/lwmd/hybo

# [ FOR FURTHER ]
# [ READING ]

Bankston, John. *Edward Teller and the Development of the Hydrogen Bomb.* Bear, Delaware: Mitchell Lane Publishers, 2002.

Cooper, Dan. *Enrico Fermi and the Revolutions of Modern Physics.* New York: Oxford University Press, 1999.

Scheibach, Michael. *Atomic Narratives and American Youth: Coming of Age with the Atom, 1945–1955.* Jefferson, NC: McFarland and Company, 2003.

# BIBLIOGRAPHY

Bankston, John. *Edward Teller and the Development of the Hydrogen Bomb.* Newark, DE: Mitchell Lane Publishers, 2002.

Cooper, Dan. *Enrico Fermi and the Revolutions of Modern Physics.* New York: Oxford University Press, 1999.

Herken, Gregg. *Brotherhood of the Bomb.* New York: Henry Holt and Company, 2002.

Moss, Norman. *Men Who Play God: The Story of the H-Bomb and How the World Learned to Live with It.* New York: Harper and Row, 1968.

Rhodes, Richard. *Dark Sun: The Making of the Hydrogen Bomb.* New York: Simon and Schuster, 1995

Roleff, Tamara L., ed. *The Atom Bomb.* San Diego: Greenhaven Press, 2000.

Teller, Edward. *Memoirs: A Twentieth-Century Journey in Science and Politics.* Boulder, CO: Perseus Publishing, 2001.

Teller, Edward. *Better a Shield than a Sword: Perspectives on the Defense and Technology.* New York: Simon & Schuster, 1987.

Winkler, Allan M. *The Cold War: A History in Documents.* New York: Oxford University Press, 2000.

Gorelik, Gennady. "The Hydrogen Bomb, 1950–1956." American Institute of Physics. Retrieved March 19, 2004 (www.aip.org/history/sakharov/hbomb.htm).

The BBC Online. "On this Day: 1954: US Tests Massive Hydrogen Bomb in Bikini." 2004. Retrieved March 19, 2004 (http://news.bbc.co.uk/onthisday/hi/dates/stories/march/1/newsid_2781000/2781419.stm).

Infoplease. "Hydrogen Bomb." 2004. Retrieved March 19, 2004 (http://www.infoplease.com/ce6/history/A0824719.html).

The Atomic Archive. "President Truman's Statement Announcing the First Soviet A-Bomb: September 23, 1949." 2004. Retrieved March 19, 2004 (http://www.atomicarchive.com/Docs/SovietAB.shtml).

Gruber, Ben. "Nuclear Blast on Bikini Atoll Still Felt 50 Years Later." March 3, 2004. Retrieved on March 19, 2004 (http://www.enn.com/news/2004-03-03/s_13655.asp).

Fox News Channel. "Hydrogen Bomb." January 29, 2003. Retrieved on April 8, 2004 (http://www.foxnews.com/printer_friendly_story/0,3566,76888,00.html).

Center for Defense Information. "Nuclear Issues: Facts at a Glance." February 4, 2003. Retrieved on April 8, 2004. (http://www.cdi.org/nuclear/facts-at-a-glance-pr.cfm).

Center for Defense Information. "Selected Nuclear Quotations: The Danger." Retrieved on April 8, 2004 (http://www.cdi.org/nuclear/nukequo.html).

# [ INDEX ]

## ABOUT THE AUTHOR

Tamra Orr is the author of more than three dozen nonfiction books for children, young adults, and families. She has a bachelor of science degree in secondary education from Ball State University in Indiana, and recently moved across the country to Portland, Oregon. She lives between the mountains and ocean and is the homeschooling mother of four children, ages eight to twenty, as well as longtime wife to Joseph.

## PHOTO CREDITS

Photo credits: cover, p. 24 © Los Alamos National Laboratory/ Science Photo Library/Photo Researchers, Inc; pp. 4–5, 26 © Time Life Pictures/Getty Images; pp. 13, 20–21, 39, 50 © Bettmann/ Corbis; p. 7 © Trinity and Beyond, VCE.com; p. 9 © American Institute of Physics/Science Photo Library/Photo Researchers, Inc.; pp. 10 (inset), 12 © Topham/The Image Works; pp. 10–11 © Soviet Group/Magnum Photos; pp. 14, 16–17, 36–37, 49 © Corbis; p. 25 © CERN/Science Photo Library/Photo Researchers, Inc.; p. 29 © Science Photo Library/Photo Researchers, Inc.; p. 31 © Photri/ Topham/The Image Works; pp. 33, 35, 43 © Hulton/Archive/ Getty Images; pp. 40–41, 54–55 © Rene Burri/Magnum Photos; pp. 44–45 © UNEP-Topham/The Image Works; p. 45 (inset) © AP/Wide World Photos/Efrem Lukatsky; p. 46 © Wally McNamee/ Corbis; p. 51 © Roger Ressmeyer/Corbis; p. 52 © Reuters/Corbis.

Designer: Evelyn Horovicz; Editor: Leigh Ann Cobb; Layout: Thomas Forget; Photo Researcher: Fernanda Rocha